KT-460-957
SERVICE

A MATERIAL WORLD

It's GLASS

KAY DAVIES and WENDY OLDFIELD

Wayland

A MATERIAL WORLD

It's Glass   It's Plastic
It's Metal   It's Wood

Editor: Joanna Housley
Designer: Loraine Hayes

First published in 1992 by
Wayland (Publishers) Ltd
61 Western Road, Hove
East Sussex BN3 1JD, England

British Library Cataloguing in Publication Data
Davies, Kay
It's Glass. – (Material World Series)
I. Title II. Oldfield, Wendy III. Series
620.144

ISBN 0 7502 0382 X

Typeset by Kalligraphic Design Ltd, Horley, Surrey
Printed and bound in Belgium by Casterman S.A.

Words that appear in **bold** in the text are explained in the glossary on page 22.

IT'S GLASS

Glass is a hard material that has been used for centuries. It is made from a very fine sand, which is heated with other ingredients until the mixture melts. When glass is heated it becomes a thick, sticky liquid. It can be shaped into different objects, which become strong and hard as the glass cools. If you look around, you will find many things that are made from glass. It is used for windows and television screens because it can be seen through. We use glasses to drink out of, and store food and drinks in glass containers, because they are strong and easy to clean. In this book you will discover some of the many different uses of glass.

The sun shines through the glass windows of the skyscraper. The windows **reflect** the world outside. Are there really clouds on the side of the building?

Pieces of coloured glass can be fitted together. They make a beautiful pattern or picture.

Have you seen **stained glass** windows like these?

Very hot glass can be blown or pulled into almost any shape we want.

When it cools the glass will become hard.

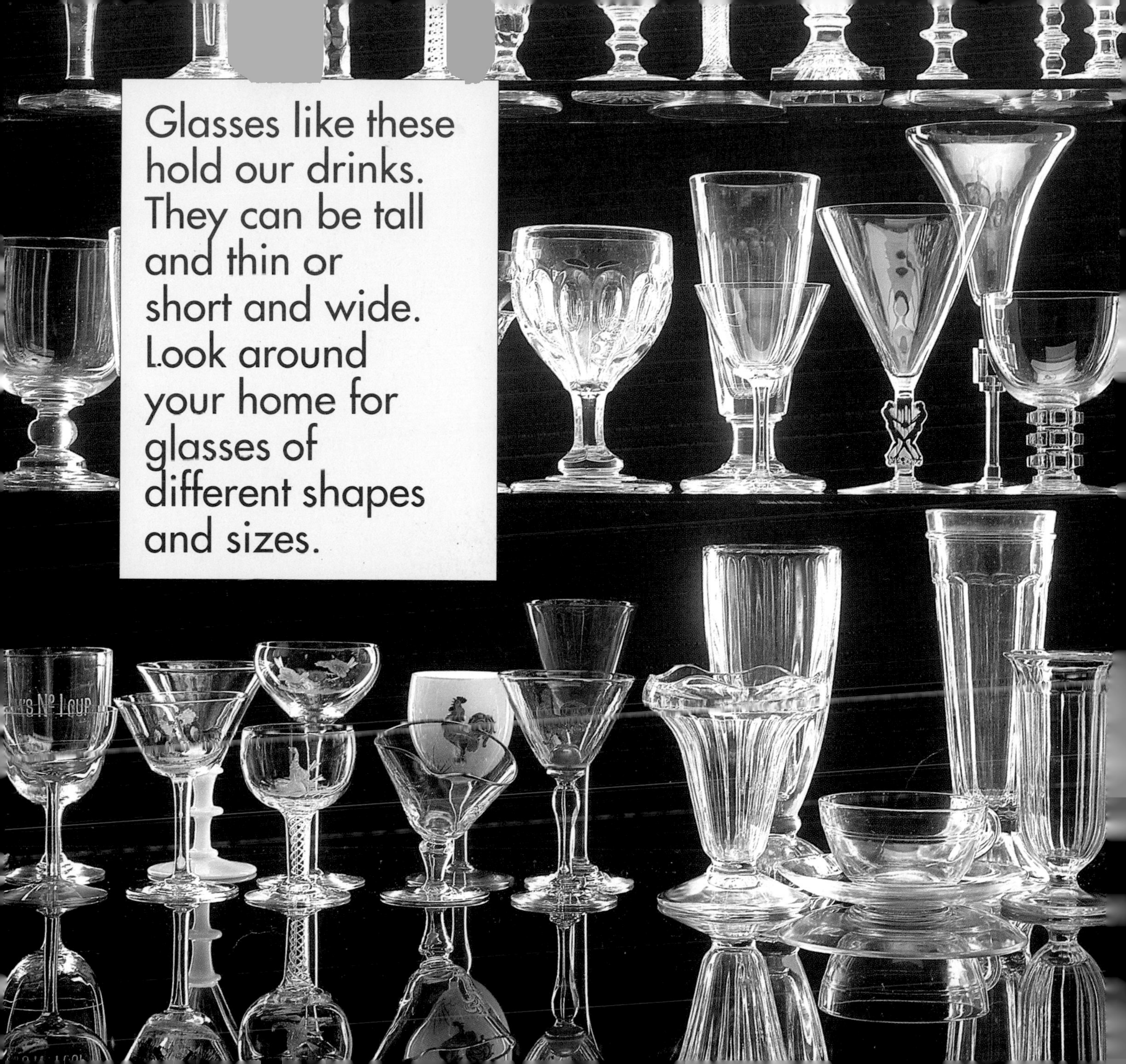

Glasses like these hold our drinks. They can be tall and thin or short and wide. Look around your home for glasses of different shapes and sizes.

Glass jars are used to store food. They keep the food fresh. When they are empty they can be washed and used again.

The water
boils in the
glass jug.
The special, strong
glass gets very
hot but it will
not break.

The coloured lights flash as the fairground rides spin. Everyone enjoys the fun of the fair.

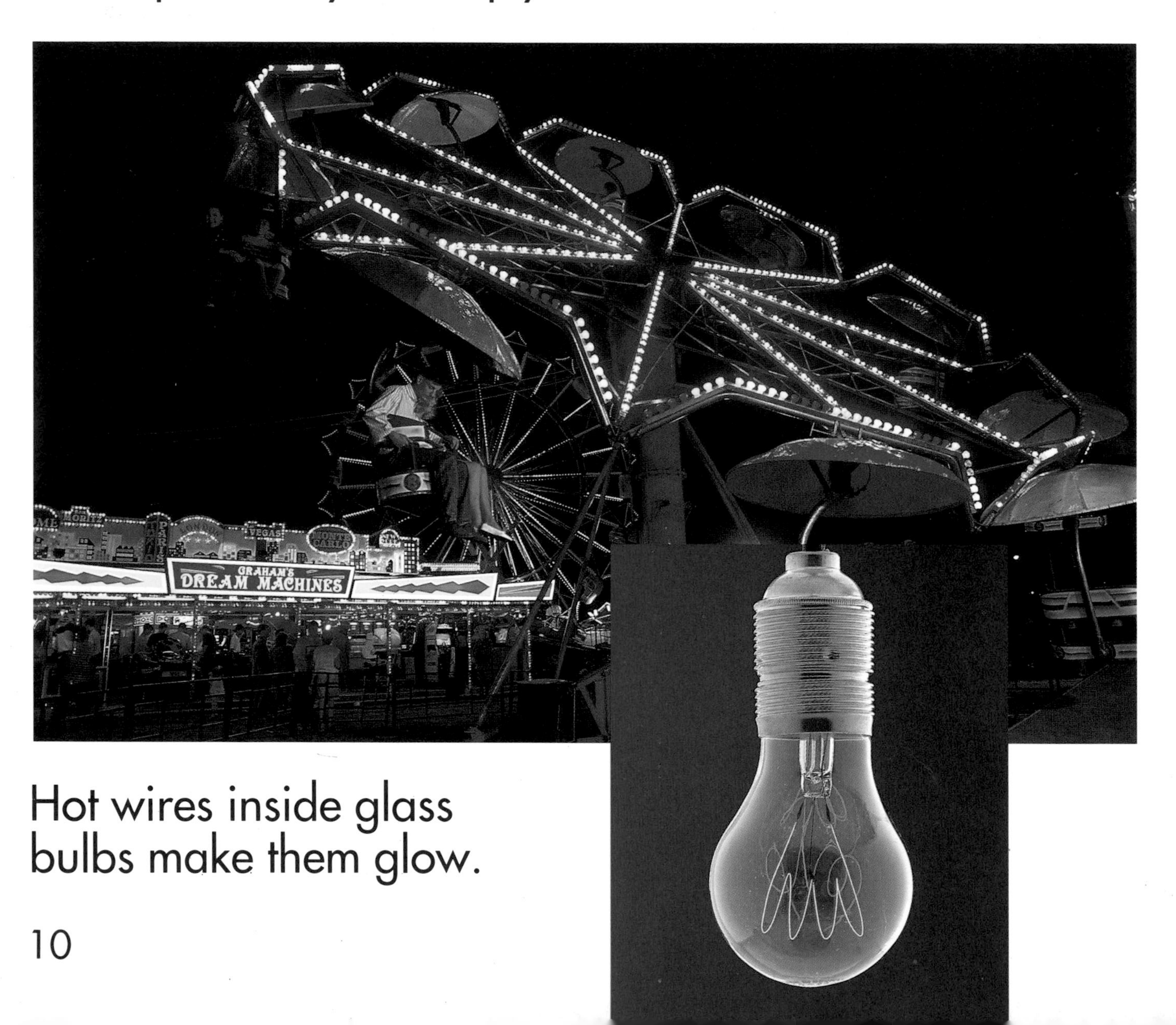

Hot wires inside glass bulbs make them glow.

The beads of glass hanging on the **chandelier** reflect the light from the candles. They fill the room with sparkling colour.

A beam of white light bends and splits into colours when it passes through a glass **prism**.

Marbles are fun to play games with. Their round, solid shape makes them strong and easy to roll across the ground. Each ball of glass has a swirl of colour inside it.

Some people wear glasses. The **lenses** are specially shaped to help people see more clearly.

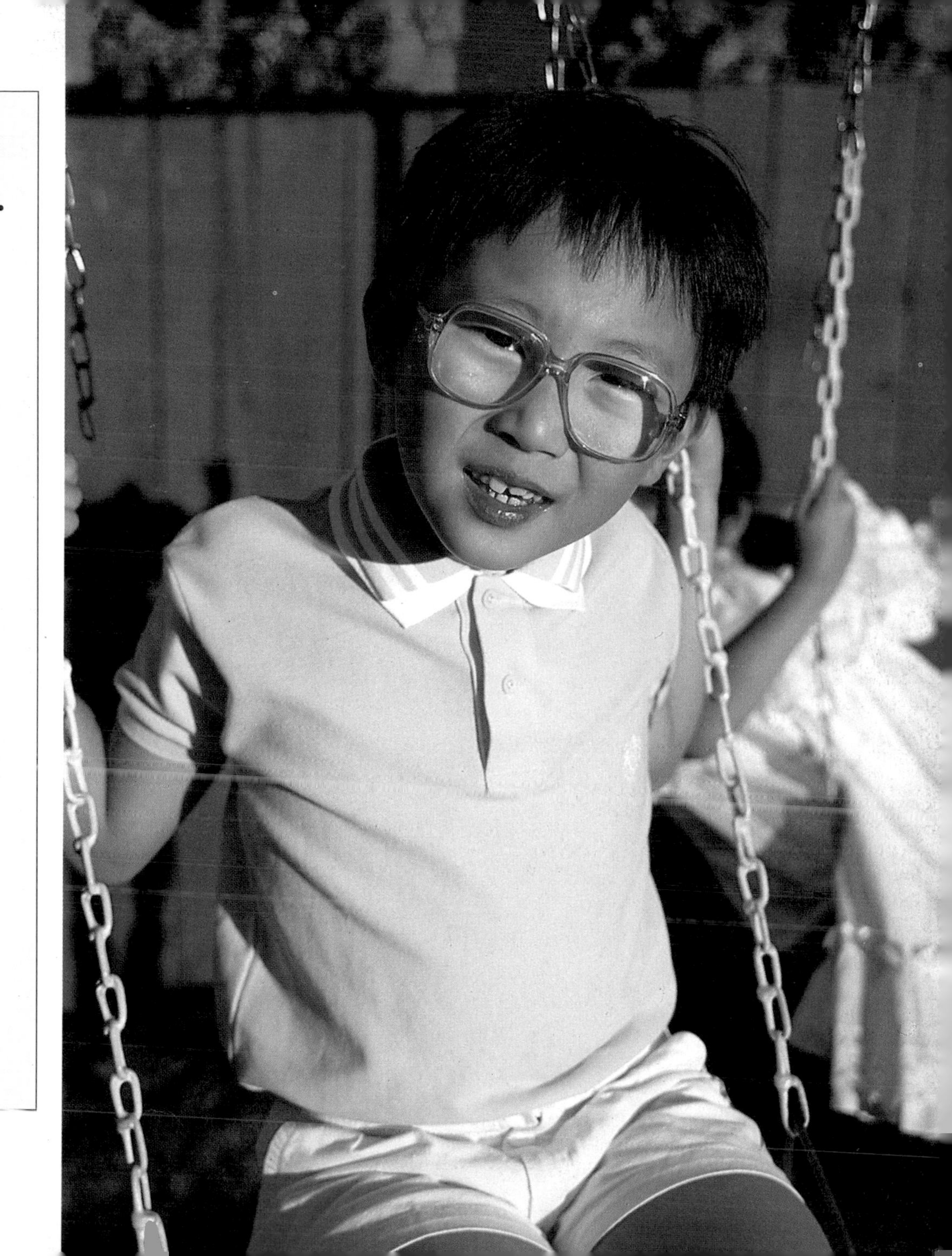

Mirrors are made from clear glass. A thin layer of metal behind the glass reflects the boy's **image**.

Glass cutters use their tools very carefully.

They can cut fine, feathery patterns which sparkle in the light.

Many creatures live in water. We can watch them twist and turn in a glass **aquarium**.

The security officer watches the televisions.

The pictures on the special glass screens show what is happening in different places inside the building.

Optical fibres are made from very thin strands of glass. Patterns of light can pass through them.

They can carry signals, such as telephone messages.

Fibreglass is a strong material made from glass. Many strands of glass are stuck and moulded together. The boat will float because fibreglass is very light.

The sharp edges of a broken bottle can cut us. We must be careful not to drop glass or knock it against anything.

Used bottles are collected in **bottle banks**. This glass will be melted and made into new objects.

GLOSSARY

Aquarium A glass tank to keep creatures in water.

Bottle bank A place to put glass for using again.

Chandelier A hanging lamp with glass decorations which reflect the light.

Fibreglass Short strands of glass that are stuck together.

Image A copy, or likeness, of something seen in a mirror.

Lens A specially shaped piece of glass to make objects appear bigger or smaller.

Optical fibres Long, thin, flexible strands of glass which carry messages in the form of light signals.

Prism A solid shape of glass which can bend white light and split it into the colours of the rainbow.

Reflect When light bounces off a smooth, shiny surface to make an image.

Stained glass Coloured pieces of glass fitted together to make a picture.

BOOKS TO READ

Materials by Kay Davies and Wendy Oldfield (Wayland, 1991)

Glass Rubbish by Veronica Bonar (Heinemann Children's Reference, 1992)

Glass by Jane Chandler (A & C Black, 1989)

Some books in A & C Black's *Simple Science* series may also be useful

TOPIC WEB

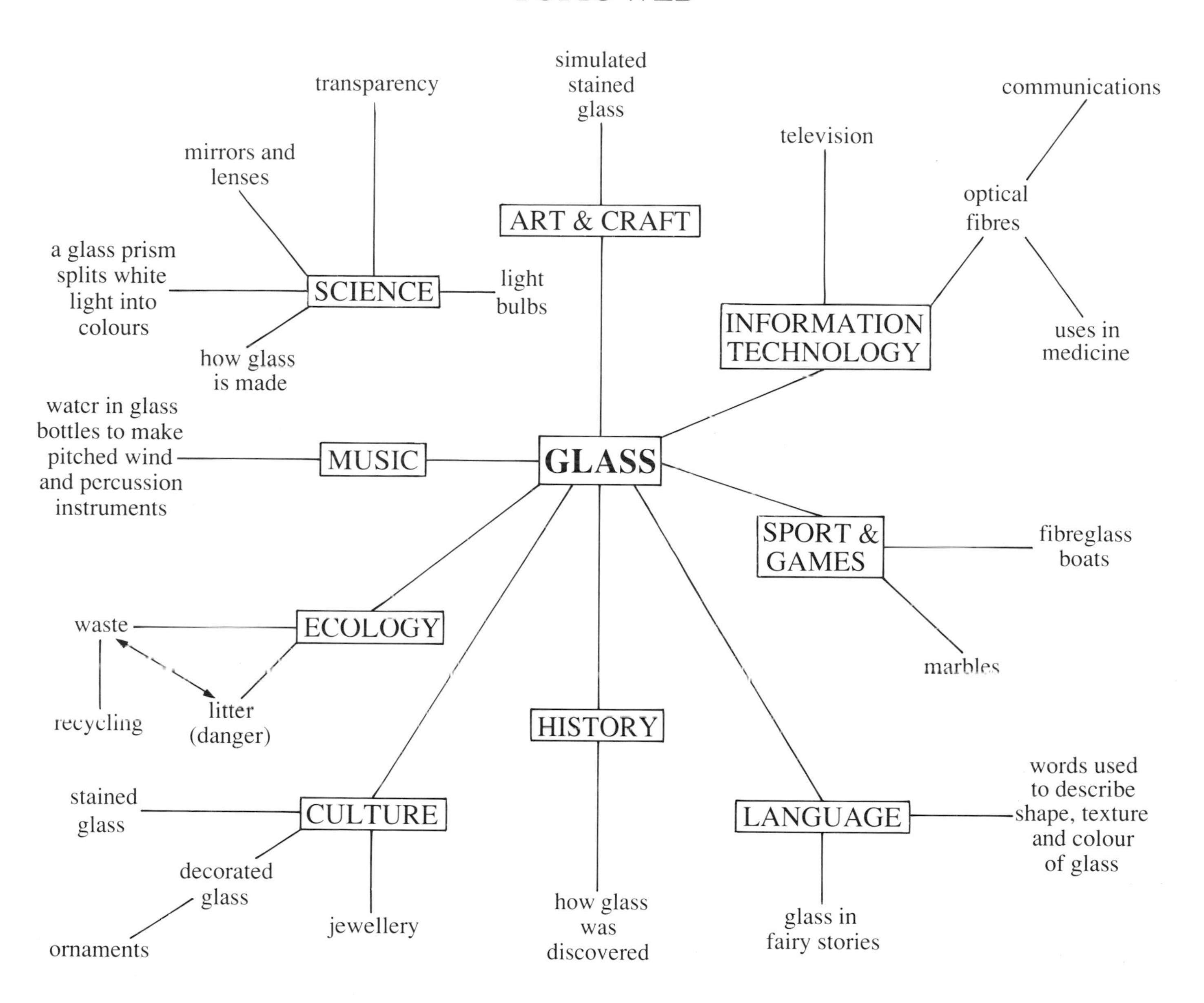

simulated stained glass
transparency
communications
television
mirrors and lenses
ART & CRAFT
optical fibres
a glass prism splits white light into colours
SCIENCE
light bulbs
INFORMATION TECHNOLOGY
uses in medicine
how glass is made
water in glass bottles to make pitched wind and percussion instruments
MUSIC
GLASS
SPORT & GAMES
fibreglass boats
waste
ECOLOGY
marbles
recycling
litter (danger)
HISTORY
stained glass
CULTURE
LANGUAGE
words used to describe shape, texture and colour of glass
decorated glass
jewellery
how glass was discovered
glass in fairy stories
ornaments

INDEX

Picture acknowledgements

Chapel Studios 5, 14, 16; Eye Ubiquitous 4 (T Baverstock), 9 (Paul Seheult), 10 (main pic); Tony Stone Worldwide 11 (main pic Peter Poulides, inset David Sutherland), 18, 19 (J A Bareham), 21 (Dave Jacobs); Topham 7, 15; Wayland Picture Library 6, 13; ZEFA 8, 10 (inset), 12 (both), 17, 20 (Paul Anthony).